Each fall and spring, the Halls go through their house. They are like a cleaning squad. First, they clean it from top to bottom. They wash windows, and they wash walls. They all use buckets of soapy water.

Next, the Halls sort all their things. They sort the old from the new. When the Halls finish, they always have a yard sale.

Wally Hall could not find anything to sell. His mother had no problem at all. She filled boxes and boxes.

Some boxes had things from when she was small. One box had old hats with veils. Some of the hats still looked quite new.

Wally's father also had boxes of things to sell. He had a few tools, which were old. He had cans of white paint. No one knew why he had them. Wally's father tossed a watch into a box. The watch did not work.

Wally wandered through his room looking for things to sell. His friends gave him a new baseball cap for his birthday. He could not sell that. He could not sell his books either. There must be something he could sell!

Wally hunted through his things again. Should he sell his cowboy hat? No, Uncle David brought that back from out west. Everything in his room had a story. How could he sell any of it?

Wally's mother called to him. "Did you find anything to sell?"

Wally looked sadly at his empty box. Just then Wally's father peeked in and said, "Wally, I think your things are just too good. If I were you, I would not sell anything."

"I can sell Mom's things," said Wally. "I can sell your things. I just can't sell my things. Do you understand?"

"I understand perfectly," said Dad.